Copycat I

<u>Second Edition</u>

**The new guide
to learning how to cook
the
most famous dishes at home**

Stephanie White

Table of contents

Introduction

Say goodbye to many hours waiting in a restaurant to have a table. Say goodbye to long lines just to buy food on tour and say goodbye to spending more than you should be eating out. Cooking at home can be time-consuming and create a mess you need to clean up, but once you prepare a particular dish and eat it, you will be proud and amazed to have made a popular snack - a delicious meal with your own hands. In this book are some good reasons why you should start cooking at home and some useful instructions to do so.

Preparing your favorite dishes at home is now easy. In this volume, you will find proven recipes from your favorite fast-food restaurant Cracker Barrel, Subway, Taco Bell, and Burger King. With these recipes, you will have the same final flavor by saving money and having fun. All the recipes here are easy to follow, and the ingredients are not hard to find! It is much cheaper to buy your ingredients for your dishes at the local farmer market. Sometimes you can even harvest it in your garden.

These recipes have been tested. I have checked repeatedly to make sure you have the correct ingredients and the correct steps to make your favorite dish. Of course, you can go online to find recipes that claim to be copies of popular restaurants. Yes, they are free. And that is why they are free. These are not real recipes for imitation purposes. It's not even close. I've tried a couple, and to be honest, they weren't close; they weren't even good. I will show you how to get the real imitation recipes and regain your freedom.

This is the best book for making imitation homemade recipes. The recipes give an accurate measurement of the ingredients, but you can modify some of them depending on how you prefer the flavor. Whether you want to minimize spice or add more to your food is up to you. You can get rid of the ingredients you are allergic to or use ingredients with lower sugar or fat content. The possibilities are endless. You have full control over that.

Chapter 1

Breakfast and Brunch

The Spinach and Artichoke Dip from Applebee's

Preparation Time: 5 minutes

Cooking Time: 30 minutes

Servings: 10

Ingredients:

- 10-ounce bag spinach, diced
- 14-ounce cans artichoke hearts, diced
- 1 cup Parmesan-Romano cheese mix, grated
- 2 cups mozzarella cheese, grated
- 16 ounces' garlic
- alfredo sauce
- 8 ounces' cream cheese, softened

Directions:

1. Combine all ingredients in a bowl. Mix well.
2. Transfer into a slow cooker. Set on high and cook for 30 minutes. Serve while hot.

Nutrition:

Calories: 228

Fat: 15 g

Carbs: 12 g

Protein: 13 g

Sodium: 418 mg

Copycat Mozzarella Sticks from TGI Fridays

Preparation Time: 10 minutes

Cooking Time: 5 minutes

Servings: 16

Ingredients:

- 2/3 cup all-purpose flour
- 2 large eggs
- 1/4cup milk
- 1 cup Japanese breadcrumbs
- 1/2cup Parmesan cheese
- 1 tablespoon dried parsley
- 1/2teaspoon garlic salt
- 1/2teaspoon seasoning salt
- 8 pieces' mozzarella string cheese
- 1-quart vegetable oil
- Marinara sauce

Directions:

1. Add flour to a bowl. Then, in a separate bowl, mix eggs and milk. Add breadcrumbs, Parmesan, parsley, garlic salt, and seasoning salt in a third bowl and mix well.
2. Line baking sheet with wax paper. Set aside.

3. Cut mozzarella pieces in half vertically so that you will end up with 16 mozzarella sticks. Then, for each piece, dredge first in flour, followed by egg wash, and third in breadcrumb mixture. Dredge again in egg wash and breadcrumbs for a thicker coat. Place pieces on prepared baking sheet and place in freezer for at least 1 hour or overnight.

4. To prepare mozzarella sticks preheat deep fryer to 350°F.

5. About 4 sticks at a time, deep fry for about 30 seconds or until golden brown. Using a slotted spoon, transfer to a rack or plate lined with paper towels to drain.

6. Serve warm with marinara sauce.

Nutrition:

Calories: 118

Fat: 7 g

Saturated Fat: 4 g

Carbs: 8 g

Sugar: 1g

Protein: 7 g

Sodium: 340 mg

The French Toasts from Denny's

Preparation Time: 10 minutes

Cooking Time: 12 minutes

Servings: 6

Ingredients:

Batter

- 4 eggs
- 2/3 cup whole milk
- 1/3 cup flour
- 1/3 cup sugar
- 1/2teaspoon vanilla extract
- 1/4teaspoon salt
- 1/8 teaspoon cinnamon

Other ingredients

- 6 slices bread loaf, sliced thick
- 3 tablespoons butter
- Powdered sugar for dusting
- Syrup as desired

Directions:

1. Mix in the ingredients for batter in a bowl.
2. Soak bread slices in batter one at a time for at least 30 seconds on both sides. Allow excess batter to drip off.
3. Melt 1 tablespoon of butter in a pan, cook battered bread over medium heat for 2 minutes or until each side is golden brown. Move slice to a plate.
4. Repeat with the remaining slices of bread, adding more butter to the pan if needed.
5. Dust with powdered sugar, if desired, and with syrup poured on top.

Nutrition:

Calories: 264

Fat: 11 g

Carbs: 33g

Protein: 8g

Sodium: 360 mg

IHOP's Healthy "Harvest Grain 'N Nut" Pancakes

Preparation Time: 5 minutes

Cooking Time: 5 minutes

Servings: 4

Ingredients:

1 teaspoon olive oil

3/4 cup oats, powdered

3/4 cup whole wheat flour

2 teaspoons baking soda

1 teaspoon baking powder

1/2teaspoon salt

11/2cup buttermilk

1/4cup vegetable oil

1 egg

1/4cup sugar

3 tablespoons almonds, finely sliced

3 tablespoons walnuts, sliced

Syrup for serving

Directions:

Heat oil in a pan over medium heat.

As pan preheats, pulverize oats in a blender until powdered. Then, add to a large bowl with flour, baking soda, baking powder and salt. Mix well.

Add buttermilk, oil, egg, and sugar in a separate bowl. Mix with an electric mixer until creamy.

Mix in wet ingredients with dry ingredients, then add nuts. Mix everything with electric mixer.

Scoop 1/3 cup of batter and cook in the hot pan for at least 2 minutes or until both sides turn golden brown. Transfer onto a plate, then repeat for the remaining batter. Serve with syrup.

Nutrition:

Calories: 433

Fat: 24 g

Carbs: 46 g

Protein: 12 g

Sodium: 1128 mg

McDonald's Sausage Egg McMuffin

Preparation Time: 10 minutes

Cooking Time: 15 minutes

Servings: 4

Ingredients:

- 4 English muffins, cut in half horizontally
- 4 slices American processed cheese
- 1/2tablespoon oil
- 1-pound ground pork, minced
- 1/2teaspoon dried sage, ground
- 1/2teaspoon dried thyme
- 1 teaspoon onion powder
- 3/4 teaspoon black pepper
- 3/4 teaspoon salt
- 1/2teaspoon white sugar

- 4 large 1/3 -inch onion ring slices
- 4 large eggs
- 2 tablespoons water

Directions:

1. Preheat oven to 300°F.
2. Cover one half of muffin with cheese, leaving one half uncovered. Transfer both halves to a baking tray. Place in oven.
3. For the sausage patties, use your hands to mix pork, sage, thyme, onion powder, pepper, salt, and sugar in a bowl. Form into 4 patties. Make sure they are slightly larger than the muffins.
4. Heat oil in a pan. Cook patties on both sides for at least 2 minutes each or until all sides turn brown. Remove tray of muffins from oven. Place cooked sausage patties on top of the cheese on muffins. Return tray to the oven.
5. In the same pan, position onion rings flat into a single layer. Crack one egg inside each of the onion rings to make them round. Add water carefully into the sides of the pan and cover. Cook for 2 minutes.
6. Remove tray of muffins from the oven. Add eggs on top of patties, then top with the other muffin half.

Serve warm.

Nutrition:

Calories: 453

Fat: 15 g

Carbs: 67 g

Protein: 15 g

Sodium: 1008 mg

Starbucks' Spinach and Feta Breakfast Wraps

Preparation Time: 5 minutes

Cooking Time: 20 minutes

Servings: 6

Ingredients:

- 10 ounces' spinach leaves
- 14 ½-ounce can dice tomatoes, drained
- 3 tablespoons cream cheese
- 10 egg whites
- 1/2teaspoon oregano
- 1/2teaspoon garlic salt
- 1/8 teaspoon pepper
- 6 whole wheat tortillas
- 4 tablespoons feta cheese, crumbled

- Cooking Spray

Directions:

1. Apply light coating of cooking spray to a pan. Cook spinach leaves on medium-high heat for 5 minutes or until leaves wilt, then stir in tomatoes and cream cheese.

2. Cook for an additional 5 minutes or until cheese is melted completely. Remove from pan and place into glass bowl and cover.

3. Set aside. In the same pan, add egg whites, oregano, salt, and pepper. Stir well and cook at least 5 minutes or until eggs are scrambled. Remove from heat.

4. Microwave tortillas for 30 seconds or until warm. Place egg whites, spinach and tomato mixture, and feta in the middle of the tortillas. Fold sides inwards, like a burrito. Serve.

Nutrition:

Calories: 157

Fat: 3 g

Carbs: 19 g

Protein: 14 g

Sodium: 305 mg

Red Lobster Blueberry Syrup

Preparation Time: 10 minutes

Cooking time: 30 minutes

 Servings: 3

Ingredients:

- 2 cups blueberries
- 1/2cup sugar - 1 cup water
- 1 tablespoon cornstarch

Directions:

1. Combine the cornstarch with 2 tablespoons of water in a small bowl. Whisk until no longer clumpy and set aside.

2. Combine the water, blueberries, and sugar in a saucepan. Bring the mixture to a boil, then reduce the heat and simmer for about 10 minutes or until it has reduced a bit. Stir in the cornstarch and whisk until well combined. Continue to simmer and stir until the sauce has thickened.

3. When it has reached a syrup-like consistency, remove from heat. You can mix with an immersion blender if you choose.

4.	Serve with pancakes or waffles.

Nutrition:

Calories 215

Protein 25

Carbs 16

Fat 5

Bacon Muffins

Preparation Time: 5 Minutes

Cooking Time: 15 Minutes

Servings: 4

Ingredients:

- 12.7oz flour
- Salt Pepper
- Egg
- 1 tsp parsley
- Four bacon pieces
- 7.8fl oz milk
- Onion
- 2 tablespoons olive oil
- 3.50ounce Cheddar cheese
- 2 tsp powder

Directions:

1. Preheat oven to 190C/170C fan forced. Line a 12-hole, 1/3 cup–capacity muffin pans with paper cases.

2. Heat oil over medium-high heat. Add bacon. Cook for 5 minutes or until crisp. Cool.

3.   Combine sifted flour with pepper, cheese, chives, and bacon in a medium bowl. Make a well in the center. Add remaining ingredients, stirring until combined.

4.   Spoon mixture into paper cases. Bake until golden and firm. Stand in pan for 5 minutes. Transfer to wire rack to cool.

Nutrition:

350 Calories

18g Fat

32g Carbohydrates

16g Protein

Breakfast Muffins

Preparation Time: 20 Minutes

Cooking Time: 20 Minutes

Servings: 2

Ingredients:

- New Thyme 1.49fl oz almond milk
- Handfuls lettuce cooked veggies
- Salt Pepper
- 1 tbsp. coriander
- 3oz granola

Directions:

1.   Line and preheat the oven at 375 degrees. Transfer and whisk the eggs in a bowl until smooth.

2.   Stir in the spinach, bacon, and cheese to the egg mixture to combine. Split the egg mixture evenly among the muffin cups. Bake until eggs are set.

3.   Serve immediately. Garnish with diced tomatoes and parsley if desired.

Nutrition:

440 calories

28g Fat

28g Carbohydrates

19g Protein

Stephanie White

Chapter 2

Appetizers and Drinks

Panda Express Toasted Ravioli

Preparation time: 10 minutes

Cooking time: 10 minutes

Servings: 4–6

Ingredients:

- 1 (1-pound) package meat ravioli, thawed
- 2 eggs
- 1/4cup water
- 1 teaspoon garlic salt
- 1 cup flour
- 1 cup breadcrumbs
- 1/2teaspoon oregano
- 1 teaspoon basil
- Grated parmesan (for garnish)

Directions:

1. Add the eggs and water to a small bowl. Beat together.
2. To another bowl, add the oregano, basil, garlic salt, and bread crumbs. Combine together.

3. To a third bowl, add the flour. Heat oil to 350°F.
4. Dip the ravioli in the flour, then the eggs, then the breadcrumbs. Repeat for each and set aside.
5. Place in oil and fry until golden brown. Place on a cooling rack or paper towel to drain some of the oil. If desired, sprinkle with parmesan. Serve with marinara sauce.

Nutrition:

Calories 228,

Total Fat 15 g,

Carbs 12 g,

Protein 13 g,

Sodium 418 mg

Red Lobster Lasagna Fritta

Preparation time: 10 minutes

Cooking time: 10 minutes

Servings: 4–6

Ingredients:

- 2/3 + 1/4cup milk (divided)
- 1 cup grated parmesan cheese, plus some more for serving
- 3/4 cup feta cheese
- 1/4teaspoon white pepper
- 1 tablespoon butter
- 7 lasagna noodles
- 1 egg
- Breadcrumbs
- Oil for frying
- 2 tablespoons marinara sauce
- Alfredo sauce, for serving

Directions:

1. Place the butter, white pepper, 2/3 cup milk, parmesan, and feta cheese in a pot. Stir and boil.
2. Prepare lasagna noodles according to instructions on the package.

3. Spread a thin layer of the cheese and milk mixture on each noodle. Fold into 2-inch pieces and place something heavy on top to keep them folded. Place in the freezer for at least 1 hour, then cut each noodle in half lengthwise.

4. In a small bowl, mix the 1/4cup milk and egg. In another bowl, place breadcrumbs.

5. Dip each piece into the egg wash then the breadcrumbs. Fry the noodles at 350°F for 4 minutes.

6. Serve by spreading some alfredo sauce at the bottom of the plate, placing the lasagna on top, and then drizzling with marinara sauce. Garnish with a sprinkle of grated parmesan cheese

Nutrition:

Calories 218,

Total Fat 15 g,

Carbs 12 g,

Protein 13 g,

Sodium 418 mg

Red Lobster Spinach Artichoke Dip

Preparation time: 15 minutes

Cooking time: 10 minutes

Servings: 4

Ingredients:

- Salt and pepper to taste
- 3 tablespoons butter
- 3 tablespoons flour
- 11/2cups milk
- 1/2teaspoon salt
- 1/4teaspoon black pepper
- 5 ounces spinach, frozen and chopped
- 1/4cup artichokes, diced (I personally like to use marinated)
- 1/2teaspoon garlic, chopped
- 1/2cup parmesan, shredded
- 1/2cup mozzarella, shredded
- 1 tablespoon asiago cheese, shredded
- 1 tablespoon romano cheese, shredded
- 2 tablespoons cream cheese
- 1/4cup mozzarella cheese (for topping)

Directions:

1. Melt butter over medium heat in a saucepan. Add flour and cook for about 1–2 minutes. Add milk and stir until thick.

2. Season with salt and pepper to taste. Add spinach, diced artichokes, garlic, cheeses, and cream cheese to the pan. Stir until warmed.

3. Pour into a small baking dish. Sprinkle mozzarella cheese on top and place under the broiler. Broil until the top begins browning.

Nutrition:

Calories 238,

Total Fat 15 g,

Carbs 12 g,

Protein 13 g,

Sodium 418 mg

Panda Express Angry Alfredo Chicken

Preparation time: 15 minutes

Cooking time: 25 minutes

Servings: 4

Ingredients:

Sauce

- 1/2cup butter
- 1 cup heavy cream
- 1/2cup freshly grated parmesan cheese
- 1/2teaspoon garlic powder
- 1/4teaspoon red pepper chili flakes

Chicken

- 1/2pound chicken breast
- Salt and pepper
- 1 tablespoon olive oil

Topping

- 1/2cup mozzarella cheese

Directions:

1. To make the sauce, heat the butter over medium-high heat in a medium-sized saucepan. Melt butter, but do not let it brown.

2. Add the heavy cream. Once it bubbles, add the cheese. Stir until thickened. Reduce heat and let simmer. Add crushed red peppers and garlic powder. Stir.

3. Season the chicken with salt and pepper. Heat olive oil over medium-high heat in a medium-sized skillet.

4. Cook chicken for 5–7 minutes on each side until cooked through. Let chicken rest, then slice into cubes.

5. Preheat oven to broil. Combine the chicken with the alfredo sauce and place in a casserole dish. Top with mozzarella cheese.

6. Place the casserole dish under the broiler. Keep in oven until the cheese begins to brown. Serve with baguette bread.

Nutrition:

Calories 228,

Total Fat 15 g,

Carbs 12 g,

Protein 13 g,

Sodium 418 mg

Panda Express Lemonade

Preparation Time: 1 minute

Cooking Time: 0 minutes

Servings: 1

Ingredients

- 1-quart of water
- 1 cup of sugar
- 1 cup of fresh lemon juice
- Sparkling water (not tonic water)

Directions

1. Mix the water, sugar, and juice.

2. Fill a glass 2/3 to 3/4 filled with the lemon mixture (depending on your preference) and top it off with soda water.

3. You can change the flavor of the lemonade by adding puréed fruit (raspberries, strawberries, etc.). You'll want to feature a touch more sugar if you're adding fresh fruit.

Nutrition:

Calories: 150

Total Fat: 0g

Carbs: 42g

Protein: 0g

Fiber: 0g

Panda Express Margarita

Preparation Time: 1 minute

Cooking Time: 0 minutes

Servings 1

Ingredients

- 11/2ounce of Cuervo or 1800 gold tequila
- 3/4 ounce of Cointreau
- 3/4 ounce of Grand Marnier
- 1/2ounce of lime juice
- 2 ounces of sour mix
- Ice, for serving

Directions

1. Refrigerate (or even freeze) the glass you plan to use.
2. While chilling, mix all the ingredients in a shaker and shake well.
3. If you wish salt on your margarita rim, pour some sea salt on a little dish, wet the rim of your chilled glass, and read the salt.
4. Add some ice, and pour the margarita mixture in.

Nutrition:

Calories: 153

Total Fat: 0g

Carbs: 7g

Protein: 0.2g

Fiber: 0.2g

Copycat Chick Fil A Frosted Coffee

Preparation Time: 15 Minutes

Cooking Time: 0 Minutes

Servings: 4

Ingredients:

- 1 cup of dark roasted coffee beans

- 2 cups of cold water

- 4 cups of Edy's Slow Churn vanilla ice cream approx. 8 measuring spoons

Directions:

1. Coarse coffee beans are grinding.

2. Place ground espresso with water in a large box and permit it to steep overnight inside the refrigerator.

3. To get rid of beans, stress via a filter cheesecloth.

4. In a mixer, upload 1 cup of coffee to two cups of ice cream.

5. Run to a milkshake's consistency. Repeat for added service

Nutrition:

Calories: 547kcal

Carbohydrate: 62 g

Protein: 9 g - Fat: 29 g

Saturated fat: 17 g - Cholesterol: 116 mg

Sodium: 213 mg - Fiber: 1 g

Potassium: 583 mg - Sugar: 56 g

White Mocha Coffee

Preparation Time: 5 Minutes

Cooking Time: 5 Minutes

Servings: 2

Ingredients:

- ¾ cup almond milk rice or your favorite milk
- ½ cup white chocolate in sparks or cut into small pieces
- 1½ - 2 cups hot coffee freshly made extra strong
- Whipped cream to taste
- White chocolate to decorate

Directions:

1. Pour the white chocolate and milk in a small pot and soften the chocolate over low heat continuously blending so that it melts less complicated and does not burn.

2. Empty the espresso into the cups, both halfway or ¾.

3. Fill what is left over with the chocolate milk combination and decorate with whipped cream and further chocolate.

4. Serve and enjoy.

Nutrition:

Calories: 82kcal

Carbohydrates: 10 g -

Protein: 1 g Fat: 4 g

Saturated fat: 2 g -

Cholesterol: 13 mg

Sodium: 60 mg

Potassium: 23 mg

Sugar: 4g

Godfather Cocktail

Preparation Time: 5 Minutes

Cooking Time: 0 Minutes

Servings: 8

Ingredients:

- 1 oz Amaretto
- 2 ounces whiskey
- 2 Luxardo Cherries ice cubes

Directions:

1. Fill a pitcher of rocks 2/three complete of ice cubes.

2. Add Amaretto then whiskey to the glass.

3. Stir well.

4. If chosen, add a few drops of Luxardo cherry syrup to the glass. Decorate with two or 3 Luxardo cherries.

Nutrition:

Calories 254 kcal

Carbs: 15

Protein: 1

Fat: 1

Sodium: 2 mg

Sugar: 13

Saturated fat: 1

Olive Garden Watermelon Moscato Sangria

Preparation Time: 5 Minutes

Cooking Time: 0 Minutes

Servings: 8

Ingredients:

- 750 MLS Moscato
- 6 6 oz Ginger Soft Drink
- 6 6 ozMonin watermelon syrup
- 4 4 cups ice
- 3/4 glass sliced strawberries
- 1 sliced orange

Directions:

1. Wash and reduce the fruit into small slices.

2. Pour Moscato right into a large jug.

3. Pour Ginger Ale and watermelon syrup into a jar. Stir gently.

4. Add ice to the jar and stir gently.

5. Add slices of strawberries and oranges.

6. Serve with slices of watermelon if desired.

Nutrition:

Calories 204

Carbs: 33

Protein: 0

Fat: 0

Saturated fat: 0

Cholesterol: 0 mg

Sodium: 19 mg

Potassium: 84 mg

Fiber: 0

Sugar: 26

Chapter 3

Pasta and Soup Recipes

Pesto Cavatappi from Noodles & Company

Preparation Time: 5 minutes

Cooking Time: 20 minutes

Servings: 8

Ingredients:

- 4 quarts' water
- 1 tablespoon salt
- 1-pound macaroni pasta
- 1 teaspoon olive oil
- 1 large tomato, finely chopped
- 4 ounce mushrooms, finely chopped
- 1/4cup chicken broth
- 1/4cup dry white wine
- 1/4cup heavy cream
- 1 cup pesto
- 1 cup Parmesan cheese, grated

Directions:

1. Add water and salt to a pot. Bring to a boil. Put in pasta and cook for 10 minutes or until al dente. Drain and set aside.

2. In a pan, heat oil. Sauté tomatoes and mushrooms for 5 minutes. Pour in broth, wine, and cream. Bring to a boil. Reduce heat to medium and simmer for 2 minutes or until mixture is thick. Stir in pesto and cook for another 2 minutes. Toss in pasta. Mix until fully coated.

3. Transfer onto plates and sprinkle with Parmesan cheese.

Nutrition:

Calories 637

Total Fat 42 g

Carbs 48 g

Protein 19 g

Sodium 1730 mg

Bob Evan's Hearty Beef Vegetable Soup

Preparation Time: 20 minutes

Cooking Time: 11 hours 10minutes

Servings: 5

Ingredients:

- 1-pound beef, lean ground
- 1 onion, chopped
- 1/2 tsp. salt
- 1/4 tsp. pepper
- 2 1/2 cups water
- 3 potatoes, and cut into cubes

- 1 can Italian diced tomatoes,
- 1 cup celery
- 1 cup carrots
- 2 tbsp. sugar
- 1 tbsp. dried parsley flakes
- 2 tsp. dried basil
- 1 bay leaf

Directions:

1. Cook onion and beef in a non-stick skillet over medium heat until meat is no longer pink, and split meat into crumbles; rinse. Season with salt and pepper.
2. Shift Slow-cooking. Add ingredients left over. Cover and simmer for 10-11 hours or until vegetables are soft. Dispose of a bay leaf before serving.

Nutrition:

Calories 218,

Total Fat 24 g,

Carbs 14 g,

Protein 17 g,

Boston Market Mac n' Cheese

Preparation Time: 10 minutes

Cooking Time: 20 minutes

Servings: 6

Ingredients:

- 1 8-ounce package spiral pasta
- 2 tablespoons butter
- 2 tablespoons all-purpose flour
- 1 3/4 cups whole milk
- 1 1/4cups diced processed cheese like Velveeta™
- 1/4teaspoon dry mustard
- 1/2teaspoon onion powder
- 1 teaspoon salt

- Pepper, to taste

Directions:

1. Cook pasta according to package instructions. Drain, then set aside.

2. To prepare sauce make the roux with four and butter over medium-low heat in a large deep skillet. Add milk and whisk until well blended. Add cheese, mustard, salt, and pepper. Keep stirring until smooth.

3. Once pasta is cooked, transfer to a serving bowl. Pour cheese mixture on top. Toss to combine.

4. Serve warm.

Nutrition:

Calories 319

Total fat 17 g

Saturated fat 10 g

Carbs 28 g

Sugar 7 g

Fibers 1 g

Protein 17 g

Sodium 1134 mg

Bob Evan's Cheddar Baked Potato Soup

Preparation Time: 10 minutes

Cooking Time: 35 minutes

Servings: 1

Ingredients:

- 1 can Campbell's Cheddar Cheese Soup
- 1 can chicken broth
- 1-pound Cheddar Cheese
- 4 cups whole milk
- 2 tbsp. butter
- 2 tbsp. Corn Starch
- 1/2 tsp. Salt
- 1/2 tsp. Pepper
- 1/2 tsp. Onion powder
- 1/2 tsp. Garlic salt
- 7 potatoes, diced and boiled

Directions:

1. Add soup, bread, 1 milk, and stir. Garnish with cheese and milk.
2. With the rest of the broth, stir in cornstarch, add to soup. Add extra spices and butter. Bring to a boil, lower heat, and cook for 15-20 minutes.

3. Add boiled potatoes, then cook for another 15 minutes. Top with peppers and chives bits' bacon. Let it cool and reheat for the best taste.

Nutrition:

Calories: 452

Fat: 19.1 g

Carbs: 15. 5 g

Protein: 23.5 g

Sodium: 250 mg

Cracker Barrel Chicken and Dumplings Soup

Preparation Time: 20 minutes

Cooking Time: 35 minutes

Servings: 4

Ingredients:

- 2 cups all-purpose flour
- 1/2 tsp. baking powder
- Salt to taste
- 2 tbsp. butter
- 1 cup milk
- 2 cans of chicken broth
- 3 cups chicken, cooked

Directions:

1. The flour, baking powder, and salt are mixed in a dish. Cut the butter with a fork or pastry blender to dry ingredients. Stir in the milk until the dough forms a ball, mixing with a fork.
2. A work surface thickly flourishes. You will need a rolling pin to cut the dumplings with and something. I like using a cutter on pizza. I like using a little spatula, too, to lift the dumplings off the cutting board.

3. With a rolling pin roll the dough thin out. Dip your cutter into the flour, and cut the dumplings into tiny squares. To them, it's all right not to be perfect. Only eye the bone. Some will be bigger, some smaller, and some could be funny in shape.

4. Put them on a heavily floured plate using the floured spatula. Just keep floating in between the dumplings layers.

5. Carry the broth to a boil to cook them. Drop the dumplings in one at a time and stir while adding them. The extra flour on them helps the broth thicken.

6. Cook them for about 20 minutes or until they taste no doughy.

7. Add the chicken to the saucepan and serve.

Nutrition:

Calories: 286

Fat: 44.1 g

Carbs: 31. 7 g

Protein: 32.3 g

Sodium: 324 mg

El Chico's Albondiga Soup

Preparation Time: 20 minutes

Cooking Time: 1 hour

Servings: 6

Ingredients

For the Meatballs (albondigas)

- 2 slices white bread
- 1/2cup milk
- 1 1/2pounds ground beef
- 1/3 cup dry long-grain rice
- 2 eggs

- 1 1/2teaspoons salt
- 2 teaspoons black pepper

For the Soup (sopa)

- 2 tablespoons vegetable oil
- 1 onion, diced
- 1 small bell pepper, diced
- 3 cloves garlic, chopped
- 3 quarts water
- 3 tomatoes, diced
- 1 cup dry rice
- 2 tablespoons salt
- 1 tablespoon cumin
- 1 tablespoon black pepper
- 3 carrots, thinly sliced
- 1 small zucchini, thinly sliced

Other Ingredients

- Cilantro
- Corn tortilla strips
- Fresh limes, cut in wedges

Directions

1. Prepare the meatballs. Preheat the oven to 400°F and line a baking tray with foil.
2. Soak the bread slices in the milk for 5–10 minutes. Add the rest of the meatball ingredients and mix to incorporate.

3. Roll the meat mixture into 1-inch balls and arrange them on the baking tray.

4. Bake for 20 minutes and drain on paper towels.

5. Prepare the soup. Heat the oil in a large soup pot or Dutch oven. Sauté the onion and pepper until they begin to sweat and soften, about 5 minutes. Add the garlic and cook 1 more minute.

6. Add the water, tomatoes, rice, and seasonings. Bring the pot to a boil, and then carefully add the meatballs, carrots, and zucchini.

7. Simmer until the rice is soft, about 30 minutes.

8. To serve, garnish with cilantro and tortilla strips. Squeeze lime juice over the bowls.

Nutrition:

Calories 221

Carbs 2 g

Fat 8.2 g

Protein 12 g

Cheesecake Factory's Pasta di Vinci

Preparation Time: 10 minutes

Cooking Time: 50 minutes

 Servings: 4

Ingredients:

- 1/2red onion, chopped
- 1 cup mushrooms, quartered
- 2 teaspoons garlic, chopped
- 1-pound chicken breast, cut into bite-size pieces
- 3 tablespoons butter, divided
- 2 tablespoons flour
- 2 teaspoons salt
- 1/4cup white wine
- 1 cup cream of chicken soup mixed with some milk
- 4 tablespoons heavy cream
- Basil leaves for serving, chopped
- Parmesan cheese for serving
- 1 pound penne pasta, cooked, drained

Directions:

1.   Sauté the onion, mushrooms and garlic in 1 tablespoon of the butter.

2.   When they are tender, remove them from the butter and place in a bowl. Cook the chicken in the same pan.

3.   When the chicken is done, transfer it to the bowl containing the garlic, onions, and mushrooms, and set everything aside.

4.   Using the same pan, make a roux using the flour and remaining butter over low to medium heat. When the roux is ready, mix in the salt, wine, and cream of chicken mixture. Continue stirring the mixture, making sure that it does not burn.

5.   When the mixture thickens, allow the mixture to simmer for a few more minutes.

6.   Mix in the ingredients that you set aside and transfer the cooked pasta to a bowl or plate.

7.   Pour the sauce over the pasta, garnish with parmesan cheese and basil, and serve.

Nutrition:

Calories 218,

Total Fat 24 g,

Carbs 14 g,

Protein 17 g,

Chapter 4

Sides and Salad

Cracker Barrel's Grits

Preparation Time: 10 minutes

Cooking time: 30 minutes

Servings: 2

Ingredients:

- 2 cups water
- 11/4cups milk
- 1 teaspoon salt
- 1 cup quick-cooking (not instant) grits
- 1/4cup butter

Directions:

1. Bring the water, milk, and salt to a boil in a small pot.
2. Whisk the grits into the liquid, stirring constantly until they are well combined.
3. Allow the mixture to return to a boil, then cover, reduce heat, and cook for about 30 minutes, stirring frequently.

4. Remove from heat and stir in the butter (and cheese, if desired).

5. Serve with butter on top.

Nutrition:

Calories 218,

Total Fat 90 g,

Carbs 66 g,

Protein 39 g,

Sodium 2038 mg

Cracker Barrel's Breaded Fried Okra

Preparation Time: 10 minutes

Cooking time: 30 minutes

Servings: 2

Ingredients:

- 1 pound fresh okra, rinsed and dried
- 1 cup self-rising cornmeal
- 1/2cup self-rising flour
- 1 teaspoon salt
- 1 cup vegetable oil (for frying)
- Salt and pepper to taste

Directions:

1. Heat the oil in a large skillet or deep fryer.
2. Cut the okra into ½-inch pieces.
3. Combine the cornmeal, flour, and salt in a large bowl.
4. Drop the okra pieces into the bowl and toss to coat. Allow to rest for a few minutes while the oil heats up.
5. Using a slotted spoon, transfer the okra from the bowl into the hot oil. Cook for about 10 minutes or until the okra has turned a nice golden color.
6. Remove from oil and place on a plate lined with paper towels to drain. Season to taste with salt and pepper.

Nutrition:

Calories 318,

Total Fat 90 g,

Carbs 66 g,

Protein 39 g,

Sodium 2038 mg

Olive Garden Pinto Beans

Preparation Time: 10 minutes

Cooking time: 60 minutes

Servings: 4

Ingredients:

- 1 pound ham hocks or country ham
- 1 tablespoon sugar
- 2 quarts water
- 2 cups dry pinto beans, sorted and washed
- 11/2teaspoons salt

Directions:

1. Cook the ham hocks until well done. Reserve the stock and pull the meat from the bone.
2. Remove any pebbles from the beans, rinse them, and add them to a large pot with the water. Season with salt and add the ham and reserved stock.
3. Bring to a boil, then reduce heat, cover and simmer for about 3 hours or until beans are tender.
4. Alternatively, you can add all of the ingredients (with the ham still on the bone) to a slow cooker and cook on low for 6–8 hours.

Nutrition:

Calories 218,

Total Fat 90 g,

Carbs 66 g,

Protein 39 g,

Sodium 203 mg

Chicken Pot stickers

Preparation Time: 40 minutes

Cooking Time: 30 minutes

Servings: 50

Ingredients:

- 1/2cup + 2 tablespoons soy sauce, divided
- 1 tablespoon rice vinegar
- 3 tablespoons chives, divided
- 1 tablespoon sesame seeds
- 1 teaspoon sriracha hot sauce
- 1-pound ground pork
- 3 cloves garlic, minced
- 1 egg, beaten
- 11/2tablespoons sesame oil
- 1 tablespoon fresh ginger, minced
- 50 dumpling wrappers
- 1 cup vegetable oil, for frying
- 1-quart water

Directions:

1. In a mixing bowl, whisk together the 1/2cup of soy sauce, vinegar, and 1 tablespoon of the chives, sesame seeds and sriracha to make the dipping sauce.

2. In a separate bowl, mix together the pork, garlic, egg, the rest of the chives, the 2 tablespoons of soy sauce, sesame oil and the ginger.

3. Add about 1 tablespoon of the filling to each dumpling wrapper. Pinch the sides of the wrappers together to seal. You may need to wet the edges a bit so that they will stick.

4. Heat the cup of oil in a large skillet. When hot, working in batches, add the dumplings and cook until golden brown on all sides. Take care of not overloading your pan.

5. Add the water and cook until tender, then serve with the dipping sauce.

Nutrition:

Calories: 140

Total Fat: 5 g

Cholesterol: 15 mg

Sodium: 470 mg

Total Carbohydrate: 19 g

Dietary Fiber: 1 g

Sugar: 2 g

Protein6 g

Cracker Barrel's Lettuce Wraps

Preparation Time: 10 minutes

Cooking Time: 10 minutes

Servings: 4

Ingredients:

- 1 tablespoon olive oil
- 2 green onions, thinly sliced
- 1-pound ground chicken
- Kosher salt and ground black pepper to taste
- 2 cloves garlic, minced
- 1 onion, diced 1/4cup hoisin sauce
- 1 tablespoon Sriracha (optional)
- 2 tablespoons soy sauce
- 1 tablespoon rice wine vinegar
- 1 tablespoon ginger, freshly grated
- 1 (8-ounce) can whole water chestnuts, diced and drained
- 1 head iceberg lettuce

Directions:

1. Add the oil to a deep skillet or saucepan and heat over medium-high heat.
2. When hot, add the chicken and cook until it is completely cooked through.
3. Stir while cooking to make sure it is properly crumbled.

4. Drain any excess fat from the skillet, then add the garlic, onion, hoisin sauce, soy sauce, ginger, sriracha and vinegar.

5. Cook until the onions have softened, then stir in the water chestnuts and green onion and cook for another minute or so. Add salt and pepper to taste. Serve with lettuce leaves and eat by wrapping them up like a taco.

Nutrition:

Calories: 157

Fat: 8 g

Cholesterol: 0 mg

Protein: 15.7 g

Carbohydrates: 10.5 g

Sugar: 2.7 g

Fiber: 1.9 g

Olive Garden Shrimp Dumplings

Preparation Time: 20 minutes

Cooking Time: 10 minutes

Servings: 6

Ingredients:

- 1-pound medium shrimp, peeled, deveined, washed, and dried, divided
- 2 tablespoons carrot, finely minced
- 2 tablespoons green onion, finely minced
- 1 teaspoon ginger, freshly minced
- 2 tablespoons oyster sauce
- 1/4teaspoon sesame oil
- 1 package wonton wrappers

Sauce:

- 1 cup soy sauce
- 2 tablespoons white vinegar
- 1/2teaspoon chili paste
- 2 tablespoons granulated sugar
- 1/2teaspoon ginger, freshly minced
- Sesame oil to taste
- 1 cup water
- 1 tablespoon cilantro leaves

Directions:

1. In a food processor or blender, finely mince 1/2pound of the shrimp.

2. Dice the other 1/2pound of shrimp.

3. In a mixing bowl, combine both the minced and diced shrimp with the remaining ingredients. Spoon about 1 teaspoon of the mixture into each wonton wrapper. Wet the edges of the wrapper with your finger, then fold up and seal tightly.

4. Cover and refrigerate for at least an hour. In a medium bowl, combine all of the ingredients for the sauce and stir until well combined.

5. When ready to serve, boil water in a saucepan and cover with a steamer. You may want to lightly oil the steamer to keep the dumplings from sticking. Steam the dumplings for 7–10 minutes.

6. Serve with sauce.

Nutrition:

Calories: 190

Total Fat: 4 g

Cholesterol: 118 mg

Sodium: 890 mg

Total Carbohydrate: 20 g

Dietary Fiber: 0 g

Sugar: 4 g

Protein: 17 g

Whole Foods® California Quinoa Salad

Preparation Time: 10 Minutes

Cooking Time: 10 Minutes

Servings: 4

Ingredients:

- 2 cups of water
- 1 cup of quinoa
- ¼ cup of balsamic vinegar
- 2 tablespoons of lime zest
- 1 mango, peeled and finely chopped
- 1 red bell pepper, finely chopped
- ½ cup of pre-cooked edamame, peeled
- 1/3 cup of red onion, diced
- ¼ cup of unsweetened coconut flakes
- ¼ cup of almonds, chopped
- ¼ cup of raisins
- 2 tablespoons of cilantro leaves, diced

Directions:

1. In a skillet, cook quinoa with water according to package's Directions.

2. Meanwhile, combine balsamic vinegar and lime zest in a bowl.

3. In a separate bowl, mix cooked quinoa, mango, bell pepper, edamame, red onion, coconut flakes, almonds, raisins, and cilantro.

4.   Add vinegar and lime zest sauce onto quinoa salad.

5.   Toss to coat salad.

6.   Serve.

Nutrition:

Calories: 351

Fat: 10 g

Carbs: 57g

Protein: 11g

Boston Market® Squash Casserole

Preparation Time: 15 Minutes

Cooking Time: 1 Hour and 20 Minutes

Servings: 8

Ingredients:

- Vegetable oil for coating
- 1 (8½-ounces) of box corn muffin mix
- 4 ½ cup of zucchini, finely chopped
- 4 ½ cup of summer squash, finely chopped
- 1/3 cup of butter
- 1½ cups of yellow onion, minced
- 1 teaspoon of salt
- ½ teaspoon of black pepper, ground
- ½ teaspoon of thyme
- 1 tablespoon of fresh parsley, sliced
- 2 chicken bouillon cubes
- 1 teaspoon of garlic, finely chopped
- 8 ounces of cheddar cheese, chopped

Directions:

1. Preheat oven to 350°F and casually coat baking tray with vegetable oil.

2. Follow the package's Directions to cook corn muffins. Set aside.

3. In a deep pan, add zucchini and summer squash.

4. Pour water into pan, enough to cover vegetables.

5. Simmer over medium-low heat or until the vegetables are soft.

6. Add the cooked squash mixture into a container along with 1 cup of the cooking water. Reserve for later. Discard remaining liquids.

7. Return pan to heat. Melt butter, and then stir-fry onions until fragrant.

8. Add salt, pepper, thyme, and parsley.

9. Stir in the chicken bouillon cubes, garlic, cooked squash and zucchini mixture, and cheese. Sprinkle with crumbled corn muffins.

10. Stir everything together until well-blended, then pour onto baking tray and cover with tinfoil.

11. Cook in oven for about 40 minutes.

12. Remove cover and bake for an extra 20 minutes.

13. Serve hot.

Nutrition:

Calories: 397;

Fat: 30g:

Carbs: 23g:

Protein: 11g

Ruth's Chris® Sweet Potato Casserole

Preparation Time: 15 Minutes

Cooking Time: 1 Hour and 20 Minutes

Servings: 6

Ingredients:

- 2 large sweet potatoes covered in aluminum foil
- 1/3 cup of plus 3 tablespoons butter, divided
- 2 tablespoons of half & half
- Salt, to taste
- ½ cup of brown sugar
- ¼ cup of all-purpose flour

Directions:

1. Preheat oven to 350°F.

2. Put sweet potatoes onto a baking tray then bake for about 60 minutes.

3. Remove from oven.

4. In a bowl, add baked sweet potatoes, 3 tablespoons butter, half & half, and salt.

5. Mash until well blended.

6. In a distinct bowl, combine pecans, brown sugar, flour, and remaining butter.

7. Transfer mashed sweet potatoes into a casserole dish, then top with pecan mixture.

8. Place in oven and bake for around 20 minutes until edges bubble and pecan topping is slightly brown.

9.   Serve.

Nutrition:

Calories: 416;

Fat: 31g - Carbs: 42g;

Fibers: 3g - Protein: 3g

Olive Garden® Salad and Creamy Dressing

Preparation Time: 5 Minutes

Cooking Time: 0 Minutes

Servings: 4

Ingredients:

Dressing:

- 1 cup of mayonnaise
- 2/3 cup of white vinegar
- 5 teaspoons of granulated sugar
- 2 tablespoons of lemon juice
- 1 tablespoon of water
- 2/3 cup of Parmesan-Romano cheese blend
- 2 teaspoons of olive oil
- 1 teaspoon of Italian seasoning
- 1 teaspoon of parsley flakes
- ½ teaspoon of garlic salt

Salad:

- 1 bag of salad blend of choice
- Red onion, sliced
- 16-20 pitted black olives, sliced
- Pepperoncini
- Roma tomato, sliced
- Croutons

- Parmesan cheese, shredded

Directions:

1. To make the dressing, add mayonnaise, white vinegar, sugar, lemon juice, water, cheese blend, olive oil, Italian seasoning, parsley, and garlic salt to a blender.

2. Pulse until well combined.

3. Store in refrigerator.

4. Ready to serve later at least 2 hours.

5. Assemble salad by layering salad blend, red onion, black olives, pepperoni, tomato, croutons, and Parmesan cheese in a bowl.

6. Serve.

Nutrition:

Calories: 127.8;

Fat: 10.2g;

Carbs: 8.9g;

Protein: 1.1g

Chapter 5

Bread and Pizzas

Olive Garden Meat Overload Pizza

Preparation Time: 25 minutes

Cooking Time: 25 minutes

Servings: 8

Ingredients

- 1 thin pizza crust, or crust of choice
- 1/2-3/4 cups marinara sauce
- 2 Tablespoons olive oil
- 1 1/2-2 pounds assorted meat like ground beef, pepperoni, Italian sausage, breakfast sausage, ham (chopped) and bacon
- Salt and pepper, to taste
- 2 cups mozzarella cheese

Directions

1. Heat oven to 425 degrees F.
2. Cook bacon until crisp. Cool slightly and then crumble.

3. Cook sausages in a little oil over medium heat to brown. Drain over paper towels.
4. Season ground beef with salt and pepper and sauté until browned. Drain.
5. Spread sauce over dough.
6. Sprinkle with about 1/2 cup mozzarella followed by half of the meat ingredients.
7. Continue layering with cheese and meat.
8. Bake until golden brown and bubbly (about 25 minutes).
9. Let set for 3-5 minutes before slicing.

Nutrition:

Calories 542

Carbs 24 g

Fat 4 g

Protein 32 g

Sodium 1685 mg

Olive Garden Classic Pepperoni

Servings: 8

Preparation Time: 15 minutes

Cooking Time: 12-15 minutes

Ingredients

- 1 thin crust pizza dough, or any dough of choice
- 1/2-3/4 basic pizza or marinara sauce
- 2 cups mozzarella, freshly shredded
- 6 ounces pepperoni

Directions

1. Preheat oven to 500 degrees F.
2. Spread sauce over crust.
3. Sprinkle with cheese.
4. Top with mozzarella.
5. Bake until golden and bubbly (about 12-15 minutes).

Nutrition:

Calories 276

Carbs 25 g

Fat 14 g

Protein 12 g

Sodium 656 mg

Olive Garden Meat with Bell Pepper & Mushrooms

Servings: 8

Preparation Time: 15 minutes

Cooking Time: 30 minutes

Ingredients

- 1 pizza crust of choice
- 1/2-3/4 cup marinara sauce
- 2 cups mozzarella, freshly shredded
- 1 1/2-2 pounds seasoned beef or pork
- 16-24 pieces pepperoni
- 1 cup mushrooms, sliced thinly
- 1 medium green bell pepper, sliced thinly
- 1 red onion, sliced

Seasoned Meat Topping:

- 2 pounds ground lean beef or pork (or combination)
- 1 teaspoon ground black pepper
- 1 teaspoon dried parsley
- 1 teaspoon oregano
- 1 teaspoon dried basil
- 1/2 teaspoon garlic powder
- 1/2 teaspoon onion powder
- 1/8 teaspoon chilli flakes

- 1/2 teaspoon paprika
- 2 teaspoons salt

Directions

1. Preheat oven to 425 degrees F.
2. Prepare the meat topping. Mix all the ingredients together well and sauté over medium heat until well-browned (about 10 minutes). Remove from heat and let cool.
3. Spread sauce over crust. and sprinkle with cheese.
4. Top with seasoned meat, pepperoni, mushrooms, bell pepper and onion.
5. Bake until golden brown (about 20 minutes).

Nutrition:

Calories 496

Carbs 27 g

Fat 30 g

Protein 27 g

Sodium 1096 mg

Panda Express Barbecue Pizza

Servings: 8

Preparation Time: 15 minutes

Cooking Time: 15 minutes

Ingredients

- 1 pizza crust of choice
- 1/2-3/4 cup barbecue sauce plus more for drizzling
- 2 cups mozzarella, freshly shredded, divided
- 1/4 cup cheddar, freshly grated
- 1 1/2-2 pounds combination of cooked smoked bacon, cooked ham and **seasoned meat**

Directions

1. Preheat oven to 500 degrees F.
2. Spread sauce over crust.
3. Sprinkle with 1 cup shredded mozzarella.
4. Add meat and top with cheddar cheese and remaining mozzarella.
5. Bake until cheese is melted and crust is golden (about 15 minutes).
6. Drizzle with barbecue sauce and let cool for 3-5 minutes to set.
7. Serve.

Nutrition:

Calories 508

Carbs 40 g

Fat 27 g

Protein 26 g

Sodium 1012 mg

Cracker Barrel Meat with Mushrooms, Bell Pepper & Olives

Servings: 8

Preparation Time: 20 minutes

Cooking Time: 20-30 minutes

Ingredients

- 1 pizza crust of choice
- 1/2 cup marinara sauce
- 2 cups mozzarella, freshly shredded
- 2 pounds combination of **seasoned meat** (pork and beef), Italian sausage, pepperoni, and ham
- 1/2 cup mushrooms, sliced thinly
- 1 medium green bell pepper, sliced into rings
- 1 red onion, sliced
- 1/4 cup black olives, pitted and sliced

Directions

1. Preheat oven to 425 degrees F.
2. Brown the meat and sausage in a little oil over medium heat until browned.
3. Slice the ham.
4. Spread sauce over crust. and sprinkle with cheese.
5. Top with seasoned meat, sausage, ham, pepperoni, mushrooms, bell pepper, onion and olives.
6. Bake until golden brown (about 20 minutes).

Nutrition:

Calories 470

Carbs 27 g

Fat 28 g

Protein 25 g

Sodium 827 mg

Panda Express Spaghetti Pizza Recipe

Preparation Time: 5 minutes Cooking Time: 30 minutes Servings: 6

Ingredients:

- 750 ml of pasta sauce
- 500 g ground beef
- 500 g of spaghetti
- 400 g of tomatoes cut into small cubes
- 150 g sliced pepperoni
- 1 1/2cups shredded cheddar cheese
- 1 cup shredded Swiss cheese
- 1/2cup grated Parmesan cheese
- 1/2cup whole milk
- 1 chopped onion
- 3 cloves garlic, minced

- 2 chopped red or green peppers
- 1 teaspoon dried Italian seasoning
- 2 large eggs

Directions:

1. Gather the ingredients to make the spaghetti pizza.
2. Preheat the oven to 170 ° C. Boil a large pot of water to cook the spaghetti.
3. Cook the beef, chopped onion, chopped garlic and chopped red and green peppers in a pan over medium heat with oil until the meat is browned.
4. Drain well and add the pasta sauce, the tomatoes cut into small cubes and the Italian seasoning. Stir well and boil over medium heat while preparing spaghetti.
5. Cook the spaghetti according to the package instructions.
6. Combine the milk, eggs and grated Parmesan cheese in a large bowl and beat until mixed.
7. Strain the spaghetti and stir with the egg mixture. Spread half of the spaghetti, egg and milk mixture in a refractory dish and copper with half of the sauce and beef mixture. Repeat the layers.
8. Bake in a preheated oven for 30 or 40 minutes until hot, and cover with the remaining cheeses and then the pepperoni. Return to the oven and bake until the cheeses melt. Let stand for five minutes and cut into squares to serve the spaghetti pizza.

Nutrition:

Calories 328,

Total Fat 24 g,

Carbs 14 g,

Protein 17 g,

Stephanie White

Chapter 6

Poultry and Fish Recipes

Cracker Barrel Chicken Fried Chicken

Preparation Time: 10 minutes

Cooking time: 30 minutes

Servings: 4

Ingredients:

Chicken

- 1/2cup all-purpose flour
- 1 teaspoon poultry seasoning
- 1/2teaspoon salt
- 1/2teaspoon pepper
- 1 egg, slightly beaten
- 1 tablespoon water
- 4 boneless skinless chicken breasts, pounded to a ½-inch thickness
- 1 cup vegetable oil

Gravy

- 2 tablespoons all-purpose flour
- 1/4teaspoon salt

- 1/4teaspoon pepper
- 11/4cups milk

Directions:

1. Preheat the oven to 200°F.
2. In a shallow dish, combine the flour, poultry seasoning, salt, and pepper.
3. In another shallow dish, mix the beaten egg and water.
4. First, dip both sides of the chicken breasts in the flour mixture, then dip them in the egg mixture, and then back into the flour mixture.
5. Heat the vegetable oil over medium-high heat in a large deep skillet. A cast iron is a good choice if you have one. Add the chicken and cook for about 15 minutes, or until fully cooked, turning over about halfway through.
6. Transfer the chicken to a cookie sheet and place in the oven to maintain temperature.
7. Remove all but 2 tablespoons of oil from the skillet you cooked the chicken in.
8. Prepare the gravy by whisking the dry gravy ingredients together in a bowl. Then whisk them into the oil in the skillet, stirring thoroughly to remove lumps. When the flour begins to brown, slowly whisk in the milk. Continue cooking and whisking for about 2 minutes or until the mixture thickens.
9. Top chicken with some of the gravy.

Nutrition:

Calories: 281

Total Fat: 30g

Carbs: 32g

Protein: 71g

Fiber: 0g

Taco Bell's Broccoli Cheddar Chicken

Preparation Time: 10 minutes

Cooking time: 45 minutes

Servings: 4

Ingredients:

- 4 skinless chicken breasts
- 1 cup milk
- 1 cup Ritz-style crackers, crushed
- 1 (10.5-ounce) can condensed cheddar cheese soup
- 1/2pound frozen broccoli
- 6 ounces cheddar cheese, shredded
- 1/2teaspoon salt
- 1/2teaspoon pepper

Directions:

1. Preheat the oven to 350°F.
2. Whisk the milk and cheddar cheese soup together in a mixing bowl.
3. Prepare a baking dish by greasing the sides, then lay the chicken in the bottom and season with the salt and pepper.
4. Pour the soup mixture over the chicken, then top with the crackers, broccoli, and shredded cheese.
5. Bake for about 45 minutes or until bubbly.

Nutrition:

Calories: 181

Total Fat: 30g

Carbs: 32g

Protein: 71g

Fiber: 0g

Grilled Chicken Tenderloin Taco Bell's

Preparation Time: 10 minutes

Cooking time: 45 minutes

Servings: 4

Ingredients:

- 4–5 boneless and skinless chicken breasts, cut into strips, or 12 chicken tenderloins, tendons removed
- 1 cup Italian dressing
- 2 teaspoons lime juice
- 4 teaspoons honey

Directions:

1. Combine the dressing, lime juice, and honey in a plastic bag. Seal and shake to combine.
2. Place the chicken in the bag. Seal and shake again, then transfer to the refrigerator for at least 1 hour. The longer it marinates, the more the flavors will infuse into the chicken.
3. When ready to prepare, transfer the chicken and the marinade to a large nonstick skillet.
4. Bring to a boil, then reduce the heat and allow to simmer until the liquid has cooked down to a glaze.

Nutrition:

Calories: 181

Total Fat: 30g

Carbs: 32g

Protein: 71g

Fiber: 0g

Taco Bell's Chicken Casserole

Preparation Time: 10 minutes

Cooking time: 60 minutes

Servings: 4

Ingredients:

Crust

- 1 cup yellow cornmeal
- 1/3 cup all-purpose flour
- 11/2teaspoons baking powder
- 1 tablespoon sugar

- 1/2teaspoon salt
- 1/2teaspoon baking soda
- 2 tablespoons vegetable oil
- 3/4 cup buttermilk
- 1 egg

Filling

- 21/2cups cooked chicken breast, cut into bite-sized pieces
- 1/4cup chopped yellow onion
- 1/2cup sliced celery
- 1 teaspoon salt
- 1/4teaspoon ground pepper
- 1 (10.5-ounce) can condensed cream of chicken soup
- 13/4 cups chicken broth
- 2 tablespoons butter
- 1/2cup melted butter

Directions:

1. Preheat the oven to 375°F.
2. To make the crust, in a large bowl, combine all of the crust ingredients until smooth.
3. Dump this mixture into a buttered or greased 8 by 8-inch baking dish. Bake for about 20 minutes, then remove from oven and allow to cool. Reduce oven temperature to 350°F.

4. Crumble the cooled cornbread mixture. Add to a large mixing bowl along with 1/2cup of melted butter. Set aside.

5. Make the chicken filling by adding the butter to a large saucepan over medium heat. Let it melt, then add the celery and onions and cook until soft.

6. Add the chicken broth, cream of chicken soup, salt, and pepper. Stir until everything is well combined. Add the cooked chicken breast pieces and stir again. Cook for 5 minutes at a low simmer.

7. Transfer the filling mixture into 4 individual greased baking dishes or a greased casserole dish. Top with the cornbread mixture and transfer to the oven.

8. Bake for 35–40 minutes for a large casserole dish or 25–30 minutes for individual dishes.

Nutrition:

Calories: 381

Total Fat: 30g

Carbs: 32g

Protein: 71g

Fiber: 0g

Cracker Barrel Sunday Chicken

Preparation Time: 10 minutes

Cooking time: 10 minutes

Servings: 4

Ingredients:

- Oil for frying
- 4 boneless, skinless chicken breasts
- 1 cups all-purpose flour
- 1 cup bread crumbs
- 2 teaspoons salt
- 2 teaspoons black pepper
- 1 cup buttermilk
- 1/2cup water

Directions:

1. Add 3–4 inches of oil to a large pot or a deep fryer and preheat to 350°F.
2. Mix the flour, breadcrumbs, salt, and pepper in a shallow dish. To a separate shallow dish, add the buttermilk and water; stir.
3. Pound the chicken breasts to a consistent size. Dry them with a paper towel, then sprinkle with salt and pepper.
4. Dip the seasoned breasts in the flour mixture, then the buttermilk mixture, then back into the flour.

5. Add the breaded chicken to the hot oil and fry for about 8 minutes. Turn the chicken as necessary so that it cooks evenly on both sides.

6. Remove the chicken to either a wire rack or a plate lined with paper towels to drain.

7. Serve with mashed potatoes or whatever sides you love.

Nutrition:

Calories: 681

Total Fat: 30g

Carbs: 32g

Protein: 71g

Fiber: 0g

Taco Bell's Creamy Chicken and Rice

Preparation Time: 10 minutes

Cooking time: 45 minutes

Servings: 4

Ingredients:

- Salt and pepper to taste
- 2 cups cooked rice
- 1 diced onion
- 1 can cream of mushroom soup
- 1 packet chicken gravy
- 11/2pounds chicken breasts, cut into strips

Directions:

1. Preheat the oven to 350°F.
2. Cook the rice. When it is just about finished, toss in the diced onion so that it cooks too.
3. Prepare a baking dish by greasing or spraying with nonstick cooking spray.
4. Dump the rice into the prepared baking dish. Layer the chicken strips on top. Spread the undiluted cream of mushroom soup over the chicken.
5. In a small bowl, whisk together the chicken gravy with 1 cup of water, making sure to get all the lumps out. Pour this over the top of the casserole.
6. Cover with foil and transfer to the oven. Bake for 45 minutes or until the chicken is completely cooked.

Nutrition:

Calories: 111

Total Fat: 23g

Carbs: 12g

Protein: 81g

Fiber: 0g

Taco Bell's Campfire Chicken

Preparation Time: 10 minutes

Cooking time: 45 minutes

Servings: 4

Ingredients:

- 1 tablespoon paprika
- 2 teaspoons onion powder
- 2 teaspoons salt
- 1 teaspoon garlic powder
- 1 teaspoon dried rosemary
- 1 teaspoon black pepper
- 1 teaspoon dried oregano
- 1 whole chicken, quartered
- 2 carrots, cut into thirds
- 3 red skin potatoes, halved
- 1 ear of corn, quartered
- 1 tablespoon olive oil
- 1 tablespoon butter
- 5 sprigs fresh thyme

Directions:

1. Preheat the oven to 400°F.
2. In a small bowl, combine the paprika, onion powder, salt, garlic powder, rosemary, pepper, and oregano.
3. Add the chicken quarters and 1 tablespoon of the spice mix to a large plastic freezer bag. Seal and refrigerate for at least 1 hour.
4. Add the corn, carrots, and potatoes to a large bowl. Drizzle with the olive oil and remaining spice mix. Stir or toss to coat.
5. Preheat a large skillet over high heat. Add some oil, and when it is hot, add the chicken pieces and cook until golden brown.
6. Lay out 4 pieces of aluminum foil and add some carrots, potatoes, corn, and a chicken quarter to each. Top with some butter and thyme.
7. Fold the foil in and make pouches by sealing the edges tightly.
8. Bake for 45 minutes.

Nutrition:

Calories: 311

Total Fat: 23g

Carbs: 12g

Protein: 81g

Fiber: 0g

Chapter 7

Beef and Pork Recipes

Southwest Steak

Preparation Time: 20 minutes

Cooking time: 10 minutes

Servings: 2

Ingredients:

- 2 (6-ounce) sirloin steaks, or your favorite cut
- 2 teaspoons blackened steak seasoning
- 1/2cup red peppers, sliced
- 1/2cup green peppers, sliced
- 2 tablespoons unsalted butter
- 1 cup yellow onion, sliced
- 2 cloves garlic, minced
- Salt, to taste
- Pepper, to taste
- 2 slices cheddar cheese
- 2 slices Monterey jack cheese

- Vegetable medley or/and garlic mashed potatoes, for serving

Directions:

1. Preheat a cast iron (or another heavy skillet) or a grill.
2. Season the meat with steak seasoning and cook to your desired doneness (about 3–4 minutes on each side for medium-rare).
3. In another skillet, melt the butter and cook the peppers, onion, and garlic. Season with salt and pepper.
4. Just before the steak has reached your desired doneness, top with a slice of each cheese and cook a bit longer until the cheese melts.
5. Serve the steaks with pepper and onion mix and garlic mashed potatoes.

Nutrition:

Calories: 350

Total Fat: 17g

Carbs: 34g

Protein: 14g Fiber: 2g

Bourbon Street Steak

Preparation Time: 10 minutes

Cooking time: 20 minutes

Servings: 4

Ingredients:

Steak ingredients

- 4 New York strip steaks, 1-inch thick
- 6 tablespoons Worcestershire sauce
- 5 tablespoons soy sauce
- 1/4cup apple cider vinegar
- 1 1/2tablespoons chili powder
- 1 1/2tablespoons garlic, minced
- 4 teaspoons meat tenderizer
- 2 tablespoons smoked paprika (regular paprika is also fine)
- 1 1/2tablespoons black pepper
- 2 teaspoons cayenne pepper
- 2 teaspoons onion salt
- 1 teaspoon dried oregano
- 1-quart beef stock

Other ingredients

- 1 tablespoon butter
- 1 onion, sliced

- 1 1/2cups sliced mushrooms
- 2 garlic cloves, minced
- 4 large potatoes, cut into 1-inch cubes
- Oil for deep frying

Directions:

1. In a large resalable bag or container with a lid, combine all the ingredients and mix to make sure they are well combined, and the steak is covered. Refrigerate for at least 8 hours or overnight, turning from time to time.
2. If you have a deep fryer, turn it on so the oil is ready to fry the potatoes. Otherwise, in a sauce pot, heat the frying oil so it reaches 350 degree F.
3. Preheat the grill, broiler, or skillet. Cook the steaks to your preference, about 4 minutes per side for medium. Transfer to a plate and keep warm with foil.
4. While steaks are cooking, in a large skillet, melt the butter over medium heat. Add the onions and sauté for 2 minutes. Add the mushrooms and sauté until mushrooms are golden, about 4-5 minutes.
5. Simultaneously, deep fry the potatoes until tender and golden brown, about 6-8 minutes. Remove from oil onto a plate or basket lined with paper towel absorb excess oil. Season with salt, and paprika, if desired.
6. Serve steaks top with mushrooms and onions and a side of potatoes.

Nutrition:

Calories: 400

Total Fat: 19g

Carbs: 39g

Protein: 16g

Fiber: 2g

DIY Sizzling Steak, Cheese, and Mushrooms Skillet

Preparation Time: 15 minutes

Cooking Time: 1 hour 35 minutes

Servings: 4

Ingredients:

- 1 head garlic, cut crosswise
- 2 tablespoons olive oil, divided
- Salt and pepper, to taste
- 2 pounds Yukon Gold potatoes, chopped into 1-inch pieces
- Water, for boiling
- 2 tablespoons butter
- 1 large yellow onion
- 8 ounces cremini mushrooms
- Salt and pepper to taste
- 1/2cup milk
- 1/4cup cream
- 3 tablespoons butter
- 21/2pounds 1-inch thick sirloin steak, cut into 4 large pieces
- 8 slices mozzarella cheese

Directions:

1. Preheat oven to 300°F.

2. Position garlic on foil. Pour 1 tablespoon olive oil to the inner sides where the garlic was cut, then wrap foil around garlic.

3. Place in oven and bake for 30 minutes. Remove from oven and squeeze out garlic from head. Transfer to a bowl or mortar. Add salt and pepper, then mash together. Set aside.

4. In a pot, add potatoes. Pour enough water on top to cover potatoes. Bring to a boil. Once boiling, reduce heat to medium. Simmer for about 20 to 25 minutes or until potatoes become tender.

5. Melt butter on a non-stick pan over medium-low heat. Add onions and sauté for about 15 minutes until a bit tender. Toss in mushrooms and sauté, adjusting heat to medium. Season with salt and pepper. Cook for 10 minutes more. Set aside and keep warm.

6. Drain potatoes, then mash using an electric mixer on low speed. While mashing, gradually pour in milk, cream, butter, and mashed garlic with olive oil. Keep blending until everything is cream-like and smooth. Remove from mixer and place a cover on top of bowl. Set aside and keep warm.

7. Evenly coat steak pieces with remaining 1 tablespoon olive oil on all sides. Heat grill, then place meat on grill.

Cook for 4 minutes. Flip and add mozzarella slices on top. Cook for another 4 minutes for medium rare. Add additional minutes for increased doneness.

8. Transfer steaks to serving plates then top with onion/mushroom mixture. Place mashed potatoes on the side. Serve.

Nutrition:

Calories: 270

Total Fat: 0g

Carbs: 0g

Protein: 22g

Fiber: 0g

PF Chang's Pepper Steak

Preparation Time: 15 minutes

Cooking time: 3 hours

Servings: 4

Ingredients:

- 11/2pounds beef sirloin
- Garlic powder to taste
- 21/2tablespoons vegetable oil
- 1 cube or 1 teaspoon beef bouillon
- 1/4cup hot water
- 1/2tablespoon cornstarch
- 1/3 cup onion, roughly chopped
- 1 green bell pepper, roughly chopped
- 1 red bell pepper, roughly chopped
- 21/2tablespoons soy sauce
- 1 teaspoon white sugar
- 1/2teaspoon salt
- 1/2teaspoon black pepper
- 1/2cup water

Directions:

1. Cut the beef into pieces approximately 11/2inches long and 1 inch wide.

2. Sprinkle the garlic powder over the beef and give it a quick stir.

3. Dissolve the bouillon in the hot water. Stir until the bouillon has completely dissolved, then stir in the cornstarch until that is completely mixed in as well.

4. Heat the oil in a large skillet or wok over medium-high heat. When hot, add the beef and vegetables. Cook just long enough to brown the beef, then transfer to a crock pot.

5. Stir the bouillon mixture a bit, then pour it over the beef in the slow cooker.

6. Add the onions, peppers, soy sauce, sugar, salt, and pepper. Add 1/2cup water around the sides of the cooker.

7. Place the cover on the slow cooker and cook for about 3 hours on high or 6 hours on low.

8. Serve with rice.

Nutrition:

Calories: 571

Total Fat: 30g

Carbs: 21g

Protein: 51g

Fiber: 2.5g

Outback Style Steak

Preparation Time: 40 minutes

Cooking time: 10 minutes

Servings: 4

Ingredients:

- 4 (6-ounce) sirloin or ribeye steaks
- 2 tablespoons olive oil
- 2 tablespoons Old Bay Seasoning
- 2 tablespoons brown sugar
- 1 teaspoon garlic powder
- 1 teaspoon salt
- 1/2teaspoon black pepper
- 1/2teaspoon onion powder
- 1/2teaspoon ground cumin

Directions:

1. Take the steaks out of the fridge and let them sit at room temperature for about 20 minutes.
2. Combine all the seasonings and mix well.
3. Rub the steaks with oil and some of the spice mixture, covering all the surfaces. Let the steaks sit for 20–30 minutes.
4. Meanwhile, heat your grill to medium-high.

5. Cook the steaks for about 5 minutes on each side for medium rare (or to an internal temperature of 130°F.) Let them sit for 5 minutes before serving.

Nutrition:

Calories: 254

Total Fat: 13g

Carbs: 56g

Protein: 45g

Fiber: 3g

Quesadilla Burger

Preparation time: 15 minutes

Cooking time: 15 minutes

Servings: 4

Ingredients:

- 1 1/2pounds ground beef 8 (6-inch) flour tortillas 1 tablespoon butter
- Tex-Mex seasoning for the burgers
- 2 teaspoons ground cumin 2 tablespoons paprika 1 teaspoon black pepper
- 1/2teaspoon cayenne pepper, more or less depending on taste
- 1 teaspoon salt or to taste 1 tablespoon dried oregano

Toppings

- 8 slices pepper jack cheese 4 slices Applewood-smoked bacon, cooked and crumbled
- 1/2cup shredded iceberg lettuce
- Pico de Galo
- 1-2 Roma tomatoes, deseeded and diced thin
- ½-1 tablespoon thinly diced onion (red or yellow is fine) 1-2 teaspoons fresh lime juice
- 1-2 teaspoons fresh cilantro, chopped finely
- 1-2 teaspoons thinly diced jalapeños pepper
- Salt and pepper to taste

Tex-Mex ranch dressing

- 1/2cup sour cream 1/2cup ranch dressing such as Hidden Valley
- 1 teaspoon Tex-Mex seasoning 1/4cup mild salsa
- Pepper to taste
- For serving (optional)
- Guacamole, and sour cream

Directions:

1. In a mixing bowl, combine the Tex-Mex seasoning ingredients and stir to ensure they are well combined.
2. Prepare the fresh Pico de Gallo by mixing all the ingredients in a bowl. Set aside in the refrigerator until ready to use.
3. Prepare the Tex-Mex ranch dressing by mixing all the ingredients in a bowl. Set aside in the refrigerator until ready to use.
4. Add 2 tablespoons of the Tex-Mex seasoning to the ground beef and mix it in, being careful not to overwork the beef or your burgers will be tough. Form into 4 large ¼-inch thick burger patties and cook either on the grill or in a skillet to your preference.
5. Heat a clean skillet over medium-low heat. Butter each of the flour tortillas on one side. Place one butter side down in the skillet.
6. Top with 1 slice of cheese, some shredded lettuce, some Pico de Gallo, some bacon, and then top with a cooked

burger. Top the burger with some of the Tex-Mex ranch dressing sauce to taste, some Pico the Gallo, bacon, and another slice of cheese.

7. Cover with another tortilla, butter side up. Cook for about 1 minute or until the tortilla is golden. Then carefully flip the tortilla and cook until the cheese has melted. This step can be done in a sandwich press if you have one. Cut the tortillas in quarters or halves and serve with a side of the Tex-Mex ranch dressing, guacamole, and sour cream, if desired.

Nutrition:

Calories: 1330,

Total Fat: 93 g,

Cholesterol: 240 mg,

Sodium: 3000 mg,

Total Carbohydrate: 50 g,

Dietary Fiber: 6 g,

Sugars: 7 g,

Protein: 74 g

Chapter 8

Dessert Recipes

Moe's Southwestern Grill's Maple Butter Blondie

Preparation Time: 15 minutes

Cooking time: 35 minutes

Servings: 9

Ingredients:

- 1/3 cup butter, melted
- 1 cup brown sugar, packed
- 1 large egg, beaten
- 1 tablespoon vanilla extract
- 1/2teaspoon baking powder
- 1/8teaspoon baking soda
- 1/8teaspoon salt
- 1 cup flour
- 2/3cup white chocolate chips
- 1/3 cup pecans, chopped (or walnuts)
- Maple butter sauce

- 3/4 cup maple syrup
- 1/2cup butter
- 3/4 cup brown sugar
- 8 ounces' cream cheese, softened to room temp
- 1/4cup pecans, chopped
- Vanilla ice cream, for serving

Directions:

1. Preheat the oven to 350°F and coat a 9x9 baking pan with cooking spray.
2. In a mixing bowl, combine the butter, brown sugar, egg, and vanilla, and beat until smooth.
3. Sift in the baking powder, baking soda, salt, and flour, and stir until it is well incorporated. Fold in the white chocolate chips.
4. Bake for 20–25 minutes.
5. While those are in the oven, prepare the maple butter sauce by combining the maple syrup and butter in a medium saucepan.
6. Cook over low heat until the butter is melted. Add the brown sugar and cream cheese. Stir constantly until the cream cheese has completely melted, then remove the pot from the heat.
7. Remove the blondies from the oven and cut them into squares.
8. Top with vanilla ice cream, maple butter sauce, and chopped nuts.

Nutrition:

Calories: 354

Fat: 12 g

Carbs: 201 g

Protein: 24.0 g

Sodium: 231 mg

Apple Chimi Cheesecake

Preparation Time: 10 minutes

Cooking time: 10 minutes

Servings: 2

Ingredients:

- 2 (9 inch) flour tortillas
- 1/4cup granulated sugar
- 1/2teaspoon cinnamon
- 3 ounces' cream cheese, softened
- 1/2teaspoon vanilla extract
- 1/3 cup apple, peeled and finely chopped
- Oil for frying
- Vanilla ice cream (optional)
- Caramel topping (optional)

Directions:

1. Make sure your tortillas and cream cheese are at room temperature; this will make them both easier to work with.
2. In a small bowl, combine the sugar and cinnamon.
3. In another mixing bowl, combine the cream cheese and vanilla until smooth. Fold in the apple.
4. Divide the apple and cheese mixture in two and place half in the center of each tortilla. Leave at least an inch margin around the outside.

5. Fold the tortilla top to the middle, then the bottom to the middle, and then roll it up from the sides.

6. Heat about half an inch of oil in a skillet over medium heat.

7. Place the filled tortillas into the skillet and fry on each side until they are golden brown. Transfer them to a paper towel lined plate to drain any excess oil, then immediately coat them with the cinnamon and sugar.

8. Serve with a scoop of ice cream.

Nutrition:

Calories: 267

Fat: 5 g

Carbs: 15 g

Protein: 18 g

Sodium: 276 mg

Chipotle's Triple Chocolate Meltdown

Preparation Time: 1 hour

Cooking time: 30 minutes

Servings: 8

Ingredients:

- 2 cups heavy cream, divided
- 1 cup white chocolate chips
- 1 cup semi-sweet chocolate chips
- 1-pound bittersweet chocolate, chopped
- 1/2cup butter, softened
- 6 eggs
- 1 1/2cups of sugar
- 1 1/2cups all-purpose flour
- Ice cream, for serving

Directions:

1. Preheat the oven to 400°F.
2. Prepare 8 ramekins by first coating the inside with butter then sprinkling them with flour so the bottom and sides are covered. Place them on a baking tray.
3. In a saucepan, bring 1 cup of heavy cream to a simmer. Remove it from the heat and add the white chocolate chips, stirring until the chocolate is melted and the mixture is smooth. Allow it to cool for about a half an hour, stirring occasionally.

4. Repeat with the other cup of cream and the semi-sweet chocolate chips.

5. In a double boiler, combine the bittersweet chocolate with the softened butter and stir until the chocolate is melted and the mixture is smooth. Remove the bowl from the heat and allow it to cool for about 10 minutes

6. In a mixing bowl, beat the eggs and the sugar together for about 2 minutes, or until the mixture is foamy. Use a rubber spatula to fold in the bittersweet chocolate mixture.

7. Turn the mixer to low and beat in the flour half a cup at a time, being careful not to overmix the batter.

8. Pour the batter evenly into the prepared ramekins and place the baking tray in oven. Bake for about 18 minutes.

9. When done, the cakes should have a slight crust but still be soft in the middle. Remove them from oven when they have reached this look. If you cook them too long you won't get the lava cake effect.

10. Let the ramekins sit on the tray for 2–3 minutes and then invert them onto serving plates.

11. Drizzle some of both the semi-sweet and white chocolate sauces over the top and serve with a scoop of ice cream.

Nutrition:

Calories: 421

Fat: 13 g

Carbs: 22 g

Protein: 24.0 g

Sodium: 311 mg

Chipotle's Chocolate Mousse Dessert Shooter

Preparation Time: 30 minutes

Cooking time: 2 minutes

Servings: 4

Ingredients:

- 2 tablespoons butter
- 6 ounces' semi-sweet chocolate chips (1 cup), divided
- 2 eggs
- 1 teaspoon vanilla
- 8 Oreo cookies
- 1/2cup prepared fudge sauce
- 2 tablespoons sugar
- 1/2cup heavy cream
- Canned whipped cream

Directions:

1. Melt the butter and all but 1 tablespoon of the chocolate chips in a double boiler.
2. When they are melted, stir in the vanilla and remove from the heat.
3. Whisk in the egg yolks.
4. Beat the egg whites until they form soft peaks, and then fold them into the chocolate mixture.

5. Beat the sugar and heavy cream in a separate bowl until it forms stiff peaks or is the consistency that you desire. Fold this into the chocolate mixture.

6. Crush the remaining chocolate chips into small pieces and stir them into the chocolate.

7. Crush the Oreos. (You can either scrape out the cream from the cookies or just crush the entire cookie.)

8. Spoon the cookie crumbs into the bottom of your cup and pat them down. Layer the chocolate mixture on top. Finish with whipped cream and either more chocolate chips or Oreo mixture.

9. Store in the refrigerator until ready to serve.

Nutrition:

Calories: 389

Fat: 11.6 g

Carbs: 25. 2 g

Protein: 39.0 g

Sodium: 222 mg

Moe's Southwestern Grill's Cinnamon Apple Turnover

Preparation Time: 10 minutes

Cooking time: 25 minutes

Servings: 6

Ingredients:

- 1 large Granny Smith apple, peeled, cored, and diced
- 1/2teaspoon cornstarch
- 1/4teaspoon cinnamon
- Dash ground nutmeg
- 1/4cup brown sugar
- 1/4cup applesauce
- 1/4teaspoon vanilla extract
- 1 tablespoon butter, melted
- 1 sheet of puff pastry, thawed
- Whipped cream or vanilla ice cream, to serve

Directions:

1. Preheat the oven to 400°F.
2. Prepare a baking sheet by spraying it with non-stick cooking spray or using a bit of oil on a paper towel.
3. In a mixing bowl, mix together the apples, cornstarch, cinnamon, nutmeg, and brown sugar. Stir to make sure

the apples are well covered with the spices. Then stir in the applesauce and the vanilla.

4. Lay out your puff pastry and cut it into squares. You should be able to make 4 or 6 depending on how big you want your turnovers to be and how big your pastry is.

5. Place some of the apple mixture in the center of each square and fold the corners of the pastry up to make a pocket. Pinch the edges together to seal. Then brush a bit of the melted butter over the top to give the turnovers that nice brown color.

6. Place the filled pastry onto the prepared baking pan and transfer to the preheated oven. Bake 20–25 minutes, or until they become a golden brown in color.

7. Serve with whipped cream or vanilla ice cream.

Nutrition:

Calories: 235

Fat: 15.8 g

Carbs: 20. 5 g

Protein: 26 g

Sodium: 109 mg

Chipotle's Cherry Chocolate Cobbler

Preparation Time: 10 minutes

Cooking time: 45 minutes

Servings: 8

Ingredients:

- 11/2cups all-purpose flour
- 1/2cup sugar
- 2 teaspoons baking powder
- 1/2teaspoon salt
- 1/4cup butter
- 6 ounces' semisweet chocolate morsels
- 1/4cup milk
- 1 egg, beaten
- 21 ounces' cherry pie filling
- 1/2cup finely chopped nuts

Directions:

1. Preheat the oven to 350°F.
2. Combine the flour, sugar, baking powder, salt and butter in a large mixing bowl. Use a pastry blender to cut the mixture until there are lumps the size of small peas.
3. Melt the chocolate morsels. Let cool for approximately 5 minutes, then add the milk and egg and mix well. Beat into the flour mixture, mixing completely.

4. Spread the pie filling in a 2-quart casserole dish. Randomly drop the chocolate batter over the filling, then sprinkle with nuts.
5. Bake for 40–45 minutes.
6. Serve with a scoop of vanilla ice cream if desired.

Nutrition:

Calories: 502

Fat: 1.8 g

Carbs: 10. 2 g

Protein: 19.0 g

Sodium: 265 mg

Cinnabon's Classic Cinnamon Rolls

Preparation Time: 2 Hours and 45 Minutes

Cooking Time: 15 Minutes

Servings: 12

Ingredients:

Dough:

- 1 cup warm milk (about 110 °F)
- 2 eggs
- 1/3 cup margarine, melted
- 4 ½ cups white bread flour
- 1 teaspoon salt
- ½ cup white sugar
- 2 ½ teaspoons rapid raising yeast
- 1 tablespoon all-purpose flour
- 1 cup brown sugar
- 2 ½ tablespoons cinnamon, ground
- 1/3 cup butter, softened

Frosting:

- 3 ounces cream cheese, softened
- ¼ cup butter, softened
- 1½ cups powdered sugar
- ½ teaspoon vanilla extract
- 1/8 teaspoon salt

Directions:

1. Arrange dough ingredients in bread machine pan following manufacturer's instructions. Select dough cycle and press Start.

2. Once the dough has reached double its original size, transfer onto surface lightly sprinkled with flour. Cover, and set aside for 10 minutes.

3. Preheat oven to 400 °F.

4. Mix brown sugar then cinnamon in a bowl.

5. Flatten dough into a rectangle, about 16 by 21 inches. Brush with 1/3 cup butter and sugar and cinnamon combination. Roll dough in a roll and cut it into 12 even pieces.

6. Handover slices onto a large baking sheet. Cover then let rest for 30 minutes or until size has doubled.

7. Bake for 15 minutes or pending lightly brown.

8. To make the frosting, combine all fixings in a bowl. Mix until smooth.

9. Remove rolls from oven and drizzle with frosting. Serve.

Nutrition:

Calories 525,

Total Fat 19 g,

Carbs 82 g,

Protein 9 g

Sodium 388 mg

Homemade Original Glazed Doughnuts from Krispy Kreme

Preparation Time: 2 Hours

Cooking Time: 30 Minutes

Servings: 4

Ingredients:

- 2 packages of rapid raising yeast (¼ ounce each)
- ¼ cup warm water, about 105°F
- 1½ cup lukewarm milk, scalded then cooled
- ½ cup white sugar
- 1 teaspoon salt
- 2 eggs
- 1/3 cup shortening
- 5 cups all-purpose flour, then more for rolling dough
- Canola oil, for deep frying

Glaze:

- 1/3 cup butter, melted
- 2 cups powdered sugar
- 1½ teaspoon vanilla
- 3 tablespoons hot water

Directions:

1. In a bowl, melt yeast in water. Then, combine with milk, ½ cup sugar, salt, eggs, shortening, and 2 cups flour.

2.	Using a mixer, whisk on low speed for 30 seconds while scraping sides of the bowl. On medium, continue to whisk for an additional 2 minutes. Mix in the rest of the flour until completely blended.

3.	Let rest for about 50 minutes or until dough rises to twice its original size.

4.	Transfer onto flat surface sprinkled with flour. Using a flour rolling pin, flatten dough until it is ½ inch thick. Cut into shape using a doughnut cutter. Let rest for another 30 minutes for the dough to rise.

5.	In a deep fryer, preheat oil to 350°F.

6.	Working in batches, carefully drop doughnuts into the deep fryer. Once floating, flip. Cook for approximately 1 minute per side or until lightly brown. Gently scoop out using a slotted spoon and transfer onto a plate lined with paper towels.

7.	Mix all the glaze ingredients, except water, in a bowl until smooth and well-blended. Gradually add water, about 1 tablespoon at a time. Whisk until smooth.

8.	Dip doughnuts in glaze. Serve.

Nutrition:

Calories 303,

Total Fat 11 g,

Carbs 48 g,

Protein 4 g,

Sodium 161 mg

Mrs. Fields Snickerdoodle Cookies

Preparation Time: 50 Minutes

Cooking Time: 14 Minutes

Servings: 16 cookies

Ingredients:

- ½ cup butter, softened
- ½ cup granulated sugar
- 1/3 cup brown sugar
- 1 egg
- ½ teaspoon vanilla
- 1½ cups flour
- ¼ teaspoon salt
- ½ teaspoon baking soda
- ¼ teaspoon cream of tartar
- 2 tablespoons granulated sugar
- 1 teaspoon cinnamon

Directions:

1.	Preheat oven to 300°F.

2.	Using a mixer, combine softened butter and sugars. Mix in egg and vanilla. Combine until there are no longer lumps.

3.	Mix flour, baking soda, salt, and cream of tartar in a bowl. Then, combine dry ingredients with wet ingredients. Blend well. Let rest in fridge for at least 30 minutes.

4.	Mix 2 tablespoons granulated sugar and teaspoon of cinnamon together in a bowl.

5.   Ball about 2½ tablespoons dough and coat evenly with cinnamon and sugar mixture. Transfer onto a baking sheet sprayed with cooking spray. Repeat for the rest of the dough.

6.   Bake for no longer than 12 minutes. Cookies should be light golden brown but still soft, not crunchy.

7.   Serve.

Nutrition:

Calories 147,

Total Fat 6 g

Carbs 22 g,

Protein 2 g

Sodium 132 mg

Cracker Barrel's Double Fudge Coca Cola Chocolate Cake

Preparation Time: 20 Minutes

Cooking Time: 40 Minutes

Servings: 12

Ingredients:

Cake:

- Non-stick cooking spray
- ½ cup unsalted butter
- ½ cup vegetable oil
- 3 tablespoons unsweetened cocoa powder
- 1 cup Coca Cola™
- 2 cups all-purpose flour
- 2 cups granulated sugar
- ½ teaspoon salt
- 1 teaspoon baking soda
- ½ cup buttermilk
- 1 teaspoon pure vanilla extract
- 2 eggs

Frosting:

- ½ cup unsalted butter (1 stick), softened
- 1 teaspoon pure vanilla extract
- 3 tablespoons unsweetened cocoa powder
- 6 tablespoons Coca Cola™

- 4 cups powdered sugar

Directions:

1.	Preheat oven to 350°F. Coat a large rectangular 9x13-inch baking pan with non-stick cooking spray.

2.	Add the butter, oil, cocoa powder, and Coca Cola™ to a saucepan. Bring to a boil. Add mixture to the electric beater bowl. Add the flour, salt, sugar, and baking powder. Beat on medium speed until well combined.

3.	Add one egg at a time. Add buttermilk and vanilla. Beat until well combined and cake batter is smooth.

4.	Transfer prepared batter into pan, spreading evenly. Place in oven then bake for 40 minutes.

5.	While the cake is in the oven, prepare the frosting. Using an electric beater, beat the butter into cream. Add 6 tablespoons of Coca Cola™, cocoa powder and vanilla. Beat until well combined.

6.	Add the powdered sugar by increment of 1 cup at a time. Beat until frosting is smooth and fluffy.

7.	Bring cake out of oven. While cake is still hot, spread the chocolate frosting evenly over the cake. Let cool down before covering with a plastic paper and place in the refrigerator until ready to serve. Serve with a scoop of vanilla ice cream, if wanted.

Nutrition:

Calories 755,

Total fat 25 g,

Saturated fat 8 g,

Carbs 108 g,

Sugar 84 g,

Fibers 2 g,

Protein 5 g,

Sodium 447 mg

Conclusion

Now that you're ready to recreate your favorite foods at home, you may want to make this a regular part of your culinary adventure. Here are some important tips and ideas to keep in mind as you go through this volume:

Take your time, and don't rush. Some recipes are quick and can be prepared in minutes, although it is best to slow down and not rush first. You can take extra time to create your first meal, to give yourself enough time. After more than once or twice, you will know exactly how long it takes to prepare your favorite dishes. Think about the time and effort you need to buy groceries, spices, etc., and how you want to use these items in your meals. Knowing the importance of good and effective use of ingredients will encourage you to organize your time and avoid wasting or overusing them at the same time.

Schedule your trips to the grocery store to avoid missing the ingredients and tapas you want with your meals. This can be done by preparing a complete shopping list, with two main categories: basic items and additives or special items. When making your shopping list, include all the basic items first, followed by any additives or special items for specific recipes or meals. This can be an easy way to keep your budget and avoid overspending, especially when it's tempting. Many of the restaurant's recipes are made from common ingredients and do not need to come from items not found in most stores or markets. For this reason, buying your common meals, as well as these recipes, should be simple and easy to manage.

When you shop, bring a list with you and a friend, neighbor, or family member so that you can both work together to choose ingredients. Preparing and buying food with others is a good tradition in some communities, which can be extended to this opportunity.

Call on your family and friends to join in the recreation of your favorite restaurants. They may have some great ideas and suggestions for preparing new foods and variations to try.

Keep a journal of your experiences in creating your favorite foods and how they will turn out. Consider improvements or ideas to change the flavor if you have a different preference. For example, some foods are spicy, although some people prefer a salty flavor and may replace cayenne or chili powder with paprika or other mild spices.

Don't expect every recipe to be perfect the first time! Sometimes the type of oven or a small change in ingredients can make a significant difference. In other cases, mistakes can be made, or a step can be skipped, although it is easy to restart the process and recreate the same dish, with improvements. Never give up, as some recipes can be a little difficult or challenging at first until you get used to the method.

If you're unsure about a particular recipe because of this, something you've never tried before (either the recipe or the food), make a small serving and buy enough for just one batch or meal. That way, you can test it yourself to see if it's worth doing again. Also, make enough samples with friends and family to see how much they like it.